Maths
Foundation
Activity Book C

T0173365

Published by Collins
An imprint of HarperCollins*Publishers*
The News Building, 1 London Bridge Street,
London, SE1 9GF, UK

HarperCollins*Publishers*
Macken House, 39/40 Mayor Street Upper,
Dublin 1, DO1 C9W8, Ireland

Browse the complete Collins catalogue at
www.collins.co.uk

© HarperCollins*Publishers* Limited 2021

10 9 8 7 6 5 4 3

ISBN 978-0-00-846879-8

British Library Cataloguing-in-Publication Data
A catalogue record for this publication is available from the British Library.

Author: Peter Clarke
Publisher: Elaine Higgleton
Product manager: Letitia Luff
Commissioning editor: Rachel Houghton
Edited by: Sally Hillyer
Editorial management: Oriel Square
Cover designer: Kevin Robbins
Cover illustrations: Jouve India Pvt Ltd.
Internal illustrations: Jouve India Pvt. Ltd.;
p 12 Tasneem Amiruddin
Typesetter: Jouve India Pvt. Ltd.
Production controller: Lyndsey Rogers
Printed and Bound in the UK using 100% Renewable
Electricity at Martins the Printers

Acknowledgements

With thanks to all the kindergarten staff and their schools around the world who
have helped with the development of this course, by sharing insights and
commenting on and testing sample materials:

Calcutta International School: Sharmila Majumdar, Mrs Pratima Nayar, Preeti
Roychoudhury, Tinku Yadav, Lakshmi Khanna, Mousumi Guha, Radhika Dhanuka,
Archana Tiwari, Urmita Das; Gateway College (Sri Lanka): Kousala Benedict; Hawar
International School: Kareen Barakat, Shahla Mohammed, Jennah Hussain; Manthan
International School: Shalini Reddy; Monterey Pre-Primary: Adina Oram; Prometheus
School: Aneesha Sahni, Deepa Nanda; Pragyanam School: Monika Sachdev; Rosary
Sisters High School: Samar Sabat, Sireen Freij, Hiba Mousa; Solitaire Global School:
Devi Nimmagadda; United Charter Schools (UCS): Tabassum Murtaza; Vietnam
Australia International School: Holly Simpson

The publishers gratefully acknowledge the permission granted to reproduce the
copyright material in this book. Every effort has been made to trace copyright
holders and to obtain their permission for the use of copyright material. The
publishers will gladly receive any information enabling them to rectify any error or
omission at the first opportunity.

Extracts from Collins Big Cat readers reprinted by permission of HarperCollins
Publishers Ltd

All © HarperCollins*Publishers*

MIX
Paper | Supporting
responsible forestry
FSC™ C007454

This book contains FSC™ certified paper
and other controlled sources to ensure
responsible forest management.

For more information visit:
www.harpercollins.co.uk/green

Trace and join

Trace the numbers. Then draw lines to join
the umbrellas in order, from 1 to 10.

Date:

How many?

| 1 | 2 | 3 | 4 | 5 | 6 | 7 | 8 | 9 | 10 |

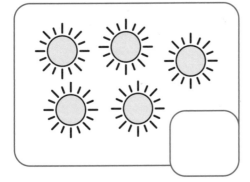

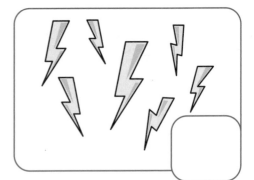

Count how many in each group.
Write the number in the box.   Date:

More

Write the number of raindrops for each cloud.
For each pair, tick the cloud with **more** raindrops.   Date:

Equal sharing

Share the apples equally between the two trees.
Draw the apples on the trees.

Date:

Take away 1

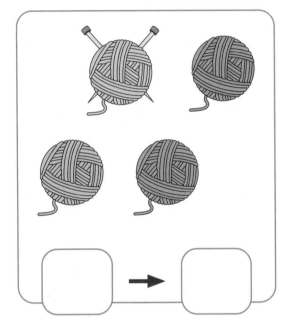

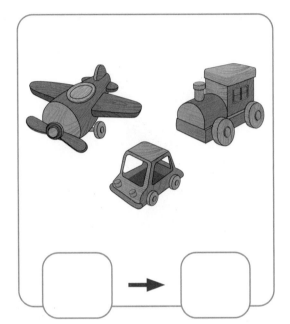

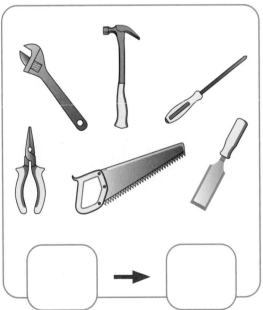

Count the objects. Record how many.
Cross out 1 object. Record how many are left.   Date:

Take away 2 or 3

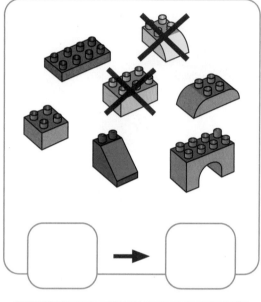

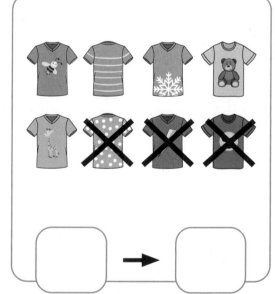

In each set, record how many there were to start with. Record how many are left.

Date:

Take away 4 or 5

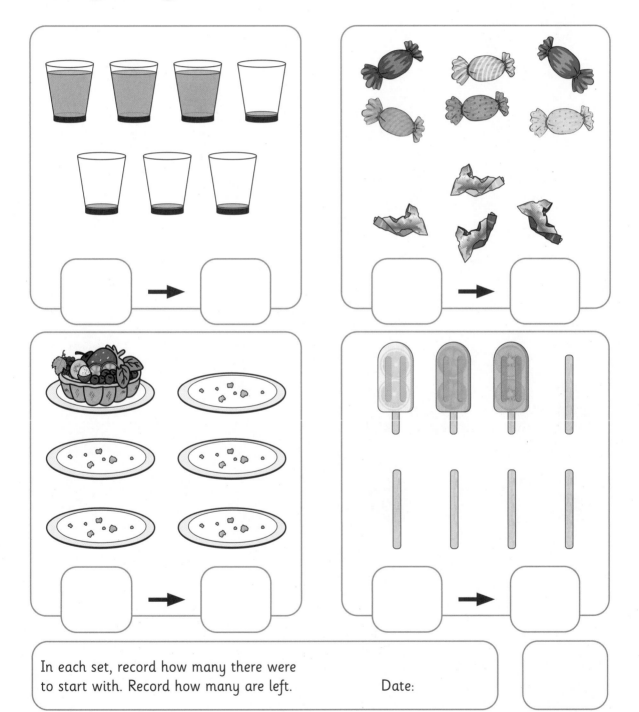

In each set, record how many there were to start with. Record how many are left.

Date:

Take away

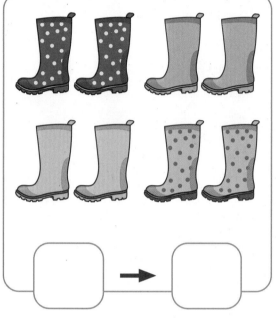

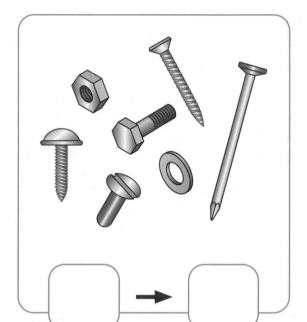

Count the objects. Record how many. Cross out up to 6 objects. Record how many are left.

Date:

I less

The frog jumps back 1 stone. Colour the stone he lands on next.

Date:

Count back 2

| 1 | 2 | 3 | 4 | 5 | 6 | 7 | 8 | 9 | 10 |

| 1 | 2 | 3 | 4 | 5 | 6 | 7 | 8 | 9 | 10 |

| 1 | 2 | 3 | 4 | 5 | 6 | 7 | 8 | 9 | 10 |

Count the seats. Circle that number on the track. Count back along the number track the number of children who are walking away. Colour the number you reach on the track.   Date:

12

Count back
3 or 4

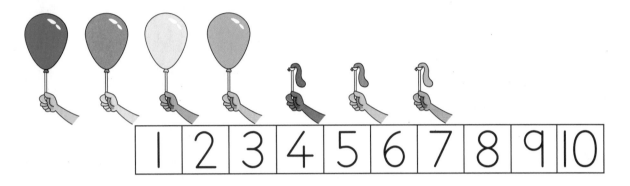

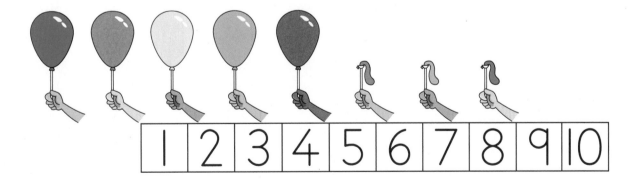

Count the sticks. Circle that number on the track. Count back along the number track the number of balloons that have burst. Colour the number you reach on the track.

Date:

Count back

| 1 | 2 | 3 | 4 | 5 | 6 | 7 | 8 | ⑨ | 10 |

| 1 | 2 | 3 | 4 | 5 | ⑥ | 7 | 8 | 9 | 10 |

| 1 | 2 | 3 | 4 | ⑤ | 6 | 7 | 8 | 9 | 10 |

| 1 | 2 | 3 | 4 | 5 | 6 | 7 | 8 | 9 | ⑩ |

Start with the circled number. Count back the number
of fingers. Colour the answer on the number track.   Date:

Heavier

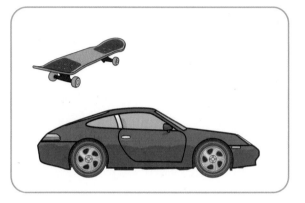

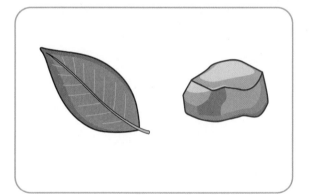

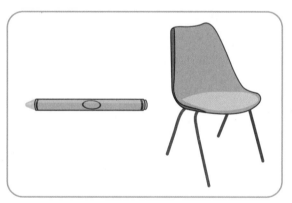

In each pair, circle the heavier object. Draw something heavier than a banana.

Date:

Balance scales

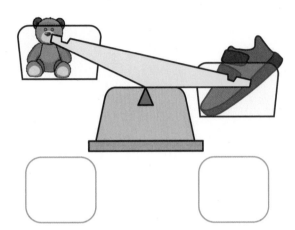

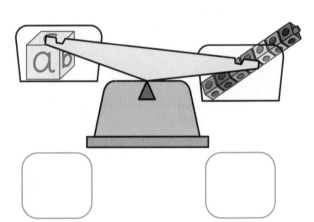

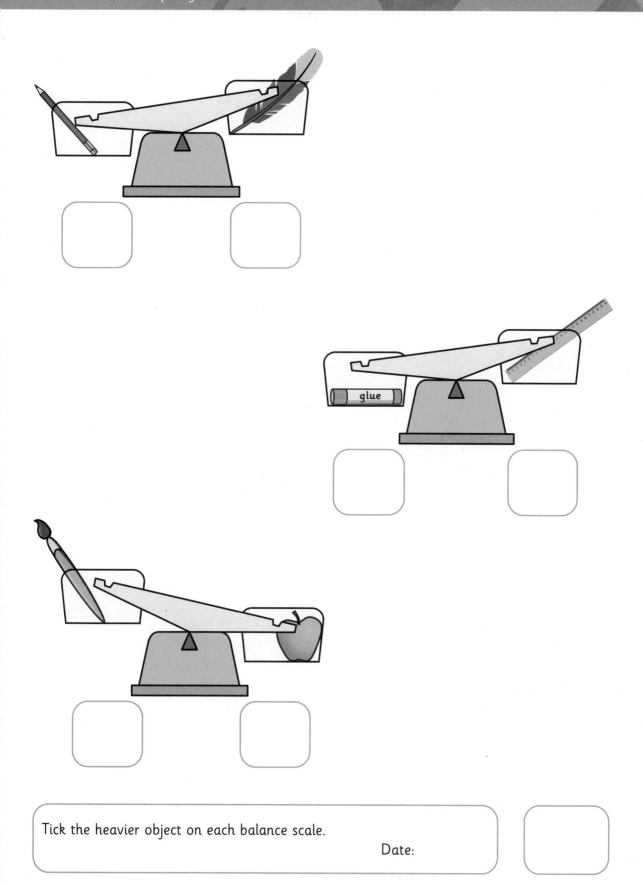

Tick the heavier object on each balance scale.

Date:

Holds more

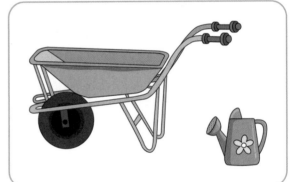

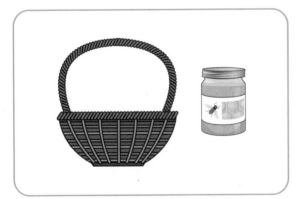

In each pair, circle the object that holds more.

Date:

Full and empty

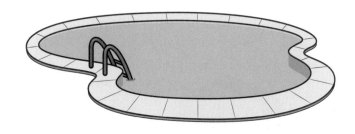

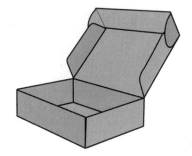

Circle in **red** all the objects that are **full**. Circle in **blue** all the objects that are **empty**.

Date:

Cubes and cuboids

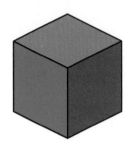

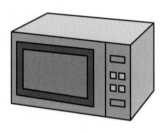

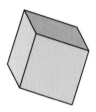

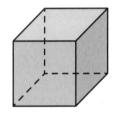

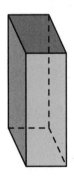

Circle in **blue** all the **cubes**. Circle in **red** all the **cuboids**.

Date:

Cylinders and spheres

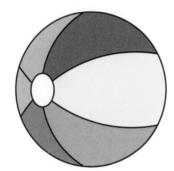

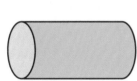

Circle in **green** all the **cylinders**. Circle in **purple** all the **spheres**.

Date:

Sort

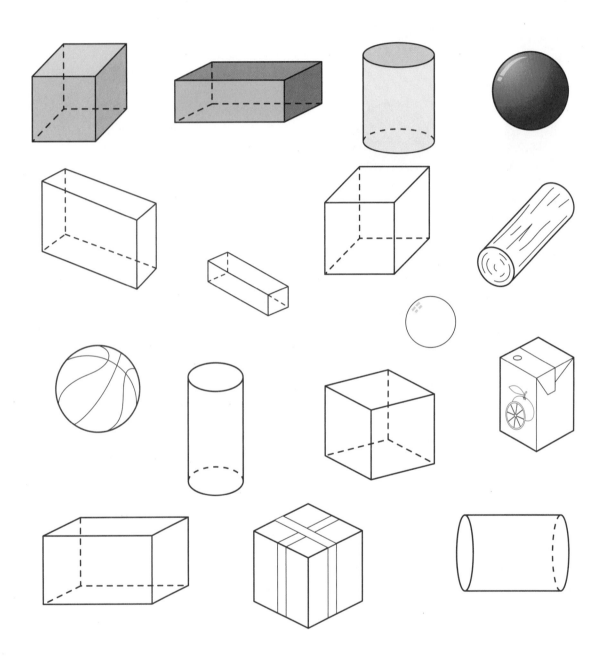

Colour each shape to match the shapes at
the top of the page.

Date:

Slide or roll

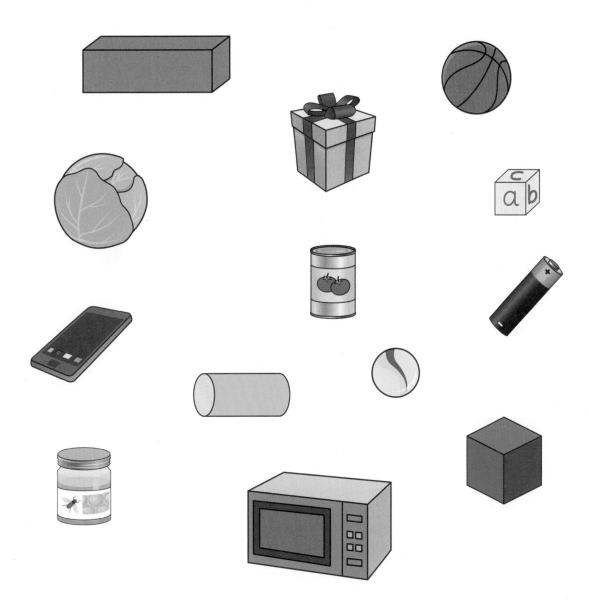

Circle all the shapes that will slide.

Date:

Assessment record

_____ has achieved these Maths Foundation Phase Objectives:

Counting and understanding numbers

- Say and use the number names in order in familiar contexts such as number rhymes, songs, stories, counting games and activities, from 1 to 10.   1   2   3
- Say the number names in order, continuing the count on or back, from 1 to 10.   1   2   3
- Count objects from 1 to 10.   1   2   3
- Count in other contexts such as sounds or actions from 1 to 10.   1   2   3
- Share objects into two equal groups.   1   2   3

Reading and writing numbers

- Recognise numbers from 1 to 10.   1   2   3
- Begin to record numbers, initially by making marks, progressing to writing numbers from 1 to 10.   1   2   3

Comparing and ordering numbers

- Use language such as more, less or fewer to compare two numbers or quantities from 1 to 10.   1   2   3

Understanding addition and subtraction

- In practical activities and discussions begin to use the vocabulary involved in subtraction: take away and counting back.   1   2   3
- Find 1 less than a number from 1 to 10.   1   2   3

Understanding shape

- Identify, describe, compare and sort 3D shapes.   1   2   3

Measurement

- Use everyday language to describe and compare mass including heavy, heavier, light and lighter.   1   2   3
- Use everyday language to describe and compare capacity and volume including more, less, full and empty.   1   2   3

Statistics

- Sort, represent and describe data using real-world objects or visual representations.   1   2   3

1: Partially achieved   2: Achieved   3: Exceeded

Signed by teacher:
Signed by parent: Date: